I0605347

HI-TECH JOBS WITHOUT COLLEGE

BE A
TELECOMMUNICATIONS TECHNICIAN

by Christopher Forest

BrightPoint Press

San Diego, CA

an imprint of ReferencePoint Press, Inc.
Printed in the United States

For more information, contact:
BrightPoint Press
PO Box 27779
San Diego, CA 92198
www.BrightPointPress.com

LIBRARY OF CONGRESS CATALOGING-IN-PUBLICATION DATA

Name: Forest, Christopher, author.
Title: Be a telecommunications technician / by Christopher Forest.
Description: San Diego, CA: BrightPoint Press, 2025 | Series: Hi-tech jobs without college | Audience: Grade 7 to 9 | Includes bibliographical references and index.
Identifiers: ISBN: 9781678209421 (hardcover) | ISBN: 9781678209438 (eBook)
The complete Library of Congress record is available at www.loc.gov.

CONTENTS

AT A GLANCE

- Telecommunications technicians work with technology that helps people and businesses communicate. This technology includes phones, television, and the internet.

- Telecommunications technicians set up equipment, lay cables, and work on communications towers.

- There are many types of telecommunications technicians. These include broadband specialists, cable installers, and broadcast engineers.

- Telecommunications technicians tend to work 40 hours a week. They may have to work unusual hours when emergencies arise.

- High schools may provide classes to help students learn skills used in telecommunications. These skills could include working with electronics.

- Technicians can get trained in many ways. They may get on-the-job training. They might work as an apprentice with an experienced technician. Or they might take a certification class.

- The demand for technicians will continue to grow. About 18,900 new jobs will open from 2022 to 2032.

- The demand for technicians will rise as technology changes. Technicians will be needed to work with 5G networks and cloud storage.

A TECHNICIAN AT WORK

A telecommunications technician named Anna parks in front of a house. She meets the homeowner at the front door. The homeowner is having trouble with her cable TV. Anna is there to solve the problem.

As a cable field technician, Anna regularly makes house calls like this one. She meets with customers. She fixes problems with their telecommunications services. This includes phone, TV, and internet equipment.

Telecommunications technicians may need to travel to several different locations in one day.

313
TACOMA
ALL TERRAIN

Some television technicians install satellite dishes, which receive TV signals from satellites.

Anna enjoys her job. Each day is different. She said, "You might run into a celebrity one day. You might be face-to-face with a skunk underneath a house the next day."[1] She likes the independence that she has while working. But she also likes the fact that she has a team that can support her if she needs help. Finally, she is grateful for the special training she undertook to become a technician.

THE PERKS OF THE JOB

Anna finds that there are many benefits to being a field technician. Her company provides her with a van and pays for gas. At the end of the day, she can take the van home. This saves her moncy. Another benefit is that she gets to help people.

Telecommunications technicians may need to bring lots of equipment to jobsites. Having a company vehicle makes this much easier.

There is room for advancement as well. In the future, she could get higher-paying jobs in the company.

There is one more benefit of the job for Anna. It is a lot of fun. She said, "I'm always laughing at my job."[2]

Telecommunications technicians, or techs, have important jobs. They look after the equipment people use to communicate over distances. They make sure this equipment keeps working. This equipment includes telephones, TV, and devices that connect to the internet. Techs are trained to make sure these services work smoothly.

CHAPTER ONE

EXPLORING TELECOMMUNICATIONS

Telecommunications, or telecom, is the practice of sending messages over distances using technology. These messages include emails, texts, and phone calls. They also include TV and radio broadcasts.

People use telecommunications devices in daily life. People's radios receive messages from radio stations. People's televisions receive signals from

The telegraph was an early telecommunications device that is rarely used today. It worked by sending signals by wire or radio. These signals could be translated into text using a code.

TV stations. Voice messages can also be sent electronically. This happens when a person makes a phone call. The internet is a newer form of technology than radio, TV, and telephones. But because the internet

People can do business faster by using telecommunications devices such as telephones than they can by using older forms of communication such as mail.

involves sending electronic messages over distances, it can be considered a form of telecom.

Individual people rely on fast communication. So do businesses. According to the company Millennia Technologies, "Telecommunication is one of the most important forms of communication in any modern business."[3] As a result, techs do important work. They make modern methods of communication possible.

Messages are sent in different ways. Radio and TV signals are sent as waves. A **transmitter** sends out these waves. They travel through the air. They arrive at a device through its **receiver**. The receiver turns the signals into sounds and images that people understand.

Messages may also be sent by cables. Some cables send information using electricity. This is the case with landline phones. Other cables, called fiber-optic cables, send information as flashes of light. These cables are commonly used for internet service.

There are two types of signals in telecom. Analog signals are one type. These signals are continuous. This means the signal can take any value within a given range. Radio signals are a type of analog signal. Digital signals are the other type of signal. These signals are not continuous.

One way to understand the difference between analog and digital signals is to picture two kinds of clocks. One kind of

Fiber-optic cables contain thin plastic or glass fibers. Light signals travel inside these fibers by reflecting from the source to the receiver.

clock is called an analog clock. It has hands that move across the clock face. As the hands move, they pass continuously through every value between the marked times. At each moment, the positions of the hands indicate the time.

The other kind of clock is called a digital clock. A digital clock displays time in certain

Both digital and analog clocks are widely used today.

intervals. A typical digital clock shows the hour and the minute. More precise clocks show the second as well. But the information presented by the clock is not continuous. The clock does not show time passing as a smooth signal.

Early telecom relied on analog signals. For example, television broadcasts were sent as analog radio waves. Today, most communication is digital. This is because digital signals are generally cheaper and more reliable than analog signals.

Internet communication uses digital signals. Cables can connect computers to the internet. These cables allow electronic messages to travel between computers in different locations. Wireless digital communication is also possible. People use routers to connect to the internet without wires. A router is a device that sends signals to computers using radio waves.

Many computers might be connected to each other at once. This happens in businesses and schools. Together, the computers form a network. People can use networks to share information. One network might be connected to other networks. The internet can be considered the network of many networks. Techs make sure internet equipment works correctly.

WHAT DO TELECOMMUNICATIONS TECHNICIANS DO?

Telecom requires a lot of equipment. That's where techs come in. They install

Emergency services, such as the 911 emergency telephone line, rely on fast telecommunications to get people help when they need it.

and maintain this equipment. This **infrastructure** includes cables and routers. It also includes transmitters and receivers.

Techs perform an important job. Their work keeps lines of communication open. Techs make it possible for people to communicate using phones and computers. They make watching TV at home possible. Techs also keep lines of communication open for emergencies. This is essential for police and fire departments.

WHAT IS THE JOB LIKE?

Being a tech involves many tasks. One important task is installing lines of communication. These lines may include wires, cables, and routers. Techs also

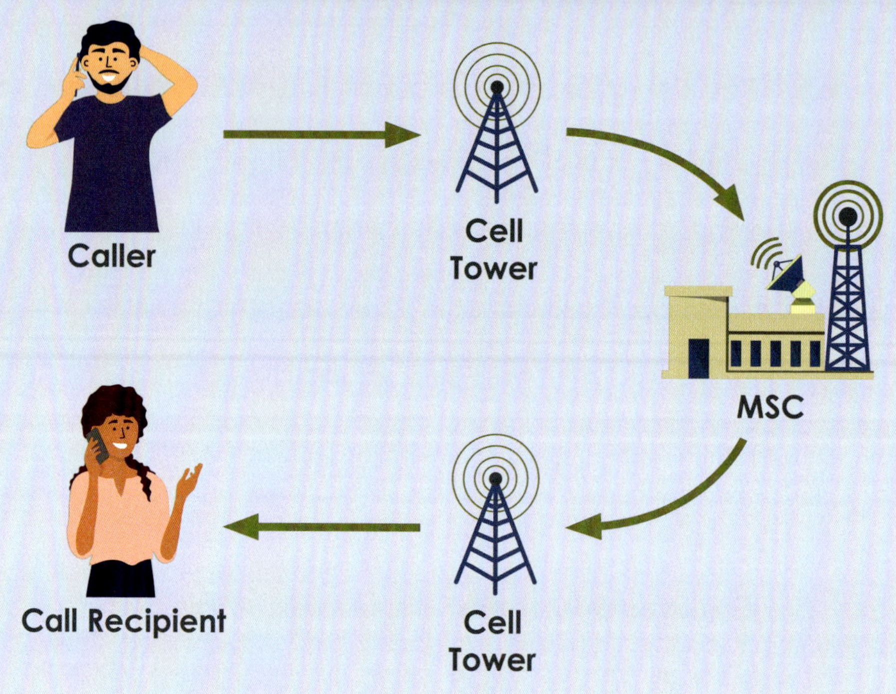

The caller's phone sends a signal to a cell tower. The tower sends the signal to a mobile switching center (MSC). This facility routes the call through a cell tower and to the recipient's phone, which begins to ring.

maintain and repair these lines. This can involve testing equipment.

In some cases, communications are broadcast by towers. Some techs work on these towers. Cell phone, television, and radio signals use towers. Cell phones emit signals to cell phone towers. Signals are sent as soon as a phone is in service. This allows the user of the phone to make a call. The call is sent out as a signal to a tower. The tower then relays the signal to the person being called. Their phone rings when the signal is received.

Some techs set up and fix phone systems. This might include installing cables and cell phone towers. It could also

Types of Technicians

There are several types of technicians. Radio techs are responsible for equipment that transmits radio waves. Broadband techs work with internet equipment. Cell phone techs maintain and repair phones.

involve working on phones themselves. Other technicians may work on computer networks. They install modems and routers. These devices allow people to access the internet. Techs might even install cable television in a person's home.

Techs may also speak with customers. They need to listen when customers talk

Many modern modems and routers are bundled in the same device.

Customer service skills are valuable for techs who interact face-to-face with customers.

about a problem. Then techs might need to explain the problem's solution. They may check in with customers about their equipment. Techs might also show customers how to use equipment.

Techs develop many skills. They become experts of the technology they work with. They learn to work well on a team. They also learn to be good problem solvers.

CHAPTER TWO

TRAINING FOR THE JOB

Being a telecom tech is a rewarding career. It is a fast-paced job that provides new challenges each day. It also gives people the chance to make a difference in the lives of customers.

There are several steps to take to become a tech. Techs have to develop the skills they need to do the job. They also need to gain experience in the field. They might do this by training on the job.

Learning from a more experienced technician can be a good way for a new tech to gain knowledge and skills.

They might also learn from a more experienced tech. Techs who want to learn more can take special courses to advance their career.

SKILLS AND EDUCATION NEEDED

Becoming a tech is hard work. Many techs develop their skills before entering the workforce. These skills can be developed in school. Techs should have strong listening skills. They often need to help customers solve problems. They should be able to understand problems the customers describe. Techs also need communication skills. They might have to explain to a customer how equipment works. Sometimes they will need to ask the customer questions. Techs should also

It's important for telecom techs to be respectful of a customer's home while working inside it.

have a flexible mindset. They may have to respond quickly to emergencies. This could mean rushing to the location of a problem at a moment's notice. Techs often need to have a driver's license as well. They typically drive to jobsites.

To get started in the field, techs need a high school diploma. A General Educational Development (GED) **credential** can be earned in place of a diploma. High school is the place where some techs become interested in the field. Many high schools offer classes in technology. Some vocational-technical schools offer telecom programs. Students in these programs

Vocational-technical schools prepare students for trade careers in fields such as construction, agriculture, and telecommunications.

spend part of their time learning skills they can use in telecom careers.

Some high schools offer internship programs in telecom. Internships match students with **professionals** in a career that interests them. Internships most often occur during a student's senior year of high school. During a telecom internship, a student could spend time with a tech. They might follow a tech who responds to service calls. Or they might work in a local cable television station to learn how television broadcasting works.

ON-THE-JOB TRAINING

Once hired, techs often get on-the-job training. A new tech might partner up with a veteran tech. The veteran tech could demonstrate the job to the new tech. New employees might also attend workshops or classes to learn specific skills. Techs are often introduced to different types of systems. These may include wireless systems and broadband systems. Techs may be trained on how to install cable lines. Techs might also learn to handle customer service.

Some companies want techs to be certified. A certification is a document that shows a tech is skilled in a particular area. Certifications can be earned by taking courses online. These courses teach skills that techs must have to work for a company. For example, a tech might take broadband certification courses. The tech would learn about internet equipment and

Some certification courses are accredited, which means that an outside organization has confirmed that the course meets certain standards.

fiber-optic cables. Completing the course earns the tech a certification.

Certifications are valuable in many ways. Some certifications allow a tech to earn more money for their work. The certifications also show customers that a company has skilled technicians. This could make customers more eager to work with that company. Certifications show future employers that a tech has special training. This makes a tech more valuable if they decide to get a new job.

Some states offer apprenticeship programs in telecom. These programs help techs learn the job by working under someone with experience. Apprentices also get paid. They may take telecom classes. These apprentices must complete training for a certain number of hours each year. For example, a telephone technician might have to take 150 hours of classes in 12 months. They might learn about telephone equipment and installation. They may take a class about telephone design. Apprenticeships can last anywhere from 1 to 6 years.

Telecom apprentices can specialize in a field of their choice, such as fiber optics or broadband services.

Being an apprentice can have several advantages. Apprentices gain a lot of job experience. Apprenticeships can connect techs with potential employers. This could help a tech get their next job faster. In 2023, Anthony Carrillo was a telecom tech apprentice in New Mexico. He was in the second year of the 3-year program. He learned by working on the job.

ApprenticeshipUSA

ApprenticeshipUSA is an organization run by the US government. It helps companies develop opportunities to train workers. These opportunities are for people who want to work in the trades, including telecom. Apprenticeships benefit workers and companies alike. They help workers begin their careers. And they help companies hire new employees.

He also took classes. He is grateful for the knowledge the program gave him. He said, “My advice for anyone thinking of joining an apprenticeship is: do it.”[4]

Becoming a telecommunications technician does not require a college degree, but it does require lots of learning.

CHAPTER THREE

WORKING IN THE FIELD

Techs often have busy schedules. They might need to perform many tasks in one day. And each day can bring something different.

Most techs work full time. This means they work an average of 40 hours a week. The hours are often flexible. Some techs need to be on call for emergencies. For example, a snowstorm might disrupt lines of communication. Techs may need to be

When poles bearing telecommunications lines go down, telecom techs are called upon to reconnect the lines.

Some companies have private phone systems. These companies rely on telecom techs to set up their systems.

prepared to work in such cases. Night and weekend shifts may be required.

Some techs work at an office every day. They might handle client calls for part of the day. Then they may have to go to a client's location. Other techs spend most of their time working out of a truck or van. They travel to different jobsites. They might install equipment at a home. Or they might solve a problem for a business. Traveling techs might have to drive hundreds of miles in a day.

The work may require techs to do physical tasks. Techs might have to climb onto roofs to install equipment. Or they may have to slide into crawl spaces to reach technology in need of repairs. Techs might also have to deal with electricity.

Average Salaries

The average pay for telecom techs was about $64,000 per year in 2023. However, pay can depend on where a tech works. In May 2023, the state that paid techs the highest average yearly wages was Rhode Island. The average was $94,400. The lowest-paying state was Arkansas. The average there was $51,230.

They should know how to do so safely in order to avoid accidents.

Much of a tech's work depends on the nature of his or her job. Different techs work with different kinds of equipment. Common types of techs include broadband

Techs who work on roofs or in other high places need to follow proper safety procedures to avoid injury.

specialists, cable installers, and broadcast engineers.

BROADBAND SPECIALIST

A broadband specialist is a tech who works with internet, phone, and cable networks. These specialists install equipment in homes, schools, and offices. They also maintain these networks. This involves running tests to make sure the networks are working. They often need to learn about the **software** used by each network. When issues arise in a network, the tech solves them. The tech might also teach customers how to use network equipment.

Eric Lagasse is a broadband specialist from North Dakota. Working with customers is an important part of the job. He teaches

The term *broadband* refers to the high volume of information that can be sent using broadband technology.

customers how to stay safe online. He also teaches them about **scams** they might encounter while using their devices. He said, "Teaching people about these issues shows we're there for them if they have questions on anything internet or security related."[5]

Broadband specialists may work in a central office. However, they may need to travel frequently. They might have to work during extreme weather events such as snowstorms or thunderstorms. These storms can cause networks to fail.

CABLE INSTALLER

A cable installer is someone who installs telecom cables. These techs mainly install fiber-optic cables. This type of cable is

Utility poles can bear power lines and telecommunications lines. Usually, telecommunications lines are strung lower than power lines.

used for telephone, television, and internet service. Cable installers also maintain the cables. They fix the cables when they need repair.

Ely Penner is a cable installer. He spends most of his time working outside. He begins his days early. He usually gets breakfast on the road while traveling to a jobsite. A job can bring extra challenges depending on where the cable needs to be installed. For example, installers might need to take boats in swampy areas to reach their destination.

Penner and his team lay the cable using machines. Some machines help pull the cable off the reel. Other machines help workers attach the cables to overhead poles. One machine keeps the cable from dropping as it is installed.

The large reels that hold telecom cables are also called drums.

Sometimes problems occur. Once, a cable became twisted on a pole while it was being laid. Penner said, "What we had to do was send up a climber so that he [could] pull it back out."[6] Then the cable could continue to be laid.

BROADCAST ENGINEER

A broadcast engineer is a special type of tech. They often work at radio or TV stations. Their role is to make sure that the station transmits its signals properly. They regularly check on broadcast signals. These techs also help maintain radio and television equipment. They understand electronics and can repair equipment as needed.

Leonard Shayo is a broadcast engineer. He works for a company that broadcasts

Program control rooms have many screens so that broadcast engineers can view video outputs from multiple cameras at once.

sports events. He spends some of his time in the program control room (PCR). This room is filled with the devices he uses during broadcasts. These devices include speakers and sound mixers. Shayo's team drives a van to events. The van carries equipment such as cameras and microphones. This equipment is used to capture sound and video of the event. This information is sent back to the PCR for broadcasting.

Broadcast engineers sometimes work around the clock when needed. They might need to respond to signal outages early in the morning. They might also need to solve interference issues. Interference happens when a radio or TV signal is disrupted.

LOOKING AHEAD

The future for techs appears bright. Techs should be in demand for the foreseeable future. One of the reasons demand is growing is that telecom is always changing. New devices and software continue to be developed. Companies will want to use these products. Telecom services will also expand into new areas. Regions that do not have reliable networks

The expansion of telecom services will be especially important for rural areas that may lack any telecom services at all.

Candidates hoping to land a new job in telecommunications need to prepare for job interviews.

will need them. Techs will be called upon to install and maintain them.

The number of job openings for techs is expected to increase in the coming years. The US Bureau of Labor Statistics estimates that the number of jobs will grow by 6 percent between 2022 and 2032. This means that 18,900 new jobs will become available in that time. That is more growth than for the average job.

Several areas in telecom are expected to see a rise in job openings. One such area is line installers and repairers. Nearly 7,500 new jobs in this area will become available by 2032. Another area of growth is equipment installers and repairers. These techs install and fix equipment such as phones and computers. About 11,000 new jobs will be available in this area by 2032.

Related Opportunities

Some techs use their job as a starting point to earn another job. They rely on the knowledge they gain to pursue other careers. For example, some techs move on to become help desk specialists at companies. They provide computer and network support for people working in their company.

Not all job openings in the telecom field will be new jobs. Some openings will be the result of current techs retiring. This will open job opportunities for new techs.

NEW TECHNOLOGY

One area of telecommunications that may not see growth is customer service jobs. There are several reasons why the demand for customer service is expected to decrease. As technology advances, it becomes more reliable. There are fewer problems to solve. Customers often become more comfortable with the tools and equipment they have over time. In recent years, artificial intelligence (AI) has been used for customer service. AI is a field of computer science that involves

Chatbots powered by artificial intelligence can save time for customer service employees.

using computers to solve problems that normally require human intelligence. Some companies are using AI chatbots that can provide customer service without the help of a real person.

AI can be useful in other ways. It can be used to make networks run smoother. It can also predict network outages and issues. This allows techs to fix issues before they cause a disaster. Navneet Sahani worked for Google **Cloud**, which offers telecom products. Sahani wrote, "Telecommunications is a fast-changing industry that is . . . hungry to learn and deploy the best possible new technologies, and that includes . . . AI."[7]

As telecom develops, the jobs of techs evolve. New products are creating new jobs. For example, the development of 5G networks for cell phones has led to a need for techs. The use of networks to store data on the cloud has also created more job opportunities for techs.

THE IMPORTANCE OF TELECOMMUNICATIONS

Telecom techs perform an important job. They install, maintain, and fix lines of communication. They make it possible for people and businesses to communicate. They allow people to watch TV, listen to the radio, and visit websites.

The work of techs is challenging. Days can be long and fast-paced. However, techs will continue to be in demand in the future. Each day on the job can bring something new and exciting. Being a telecom tech can be very rewarding.

As long as people rely on technology to communicate quickly over long distances, telecom techs will be in demand.

GLOSSARY

cloud

a network of servers that can be accessed over the internet

credential

a qualification or achievement that shows a person is capable of doing a specific job or task

infrastructure

the basic systems and structures that support a larger body, such as a company or a city

professionals

paid workers who have studied and trained to do a job properly

receiver

a device that gathers an audio or visual signal

scams

plans meant to trick or fool a person, often to take advantage of the person

software

the programs that a computer uses

transmitter

a device used to make and send a digital or audio signal

SOURCE NOTES

INTRODUCTION: A TECHNICIAN AT WORK

1. Quoted in Charter Communications, "A Day in the Life of a Field Tech," *YouTube*, June 5, 2017. www.youtube.com.

2. Quoted in Charter Communications, "A Day in the Life of a Field Tech."

CHAPTER ONE: EXPLORING TELECOMMUNICATIONS

3. "The Benefits & Importance of Using Telecommunication," *Millennia Technologies* (blog), *Millennia Technologies*, n.d. www.mtvoip.com.

CHAPTER TWO: TRAINING FOR THE JOB

4. Quoted in NMDWS, "Telecommunications Apprenticeship Program," *YouTube*, February 6, 2023. www.youtube.com.

CHAPTER THREE: WORKING IN THE FIELD

5. Quoted in "The Fascinating Day in the Life of a Telecom Professional," *United Communications* (blog), *United and Turtle Mountain Communications*, June 5, 2023. www.utma.com.

6. Ely Penner, "A Day on My Job | Installing a Fiber Optic," *YouTube*, July 24, 2021. www.youtube.com.

CHAPTER FOUR: LOOKING AHEAD

7. Navneet Sahani, "Generative AI: The Next Phase of Cloud Transformation for Communications Service Providers," *Google Cloud Blog*, *Google Cloud*, June 15, 2023. https://cloud.google.com.

FOR FURTHER RESEARCH

BOOKS

Cynthia Kennedy Henzel, *Be a Cybersecurity Specialist*. San Diego, CA: BrightPoint Press, 2025.

Blake Hoena, *Cell Phones and Smartphones: A Graphic History*. Minneapolis, MN: Graphic Universe, 2021.

Emmett Martin, *Wi-Fi: Connecting Us Online*. New York: Gareth Stevens, 2023.

INTERNET SOURCES

"Career Exploration and Skill Development," *Youth.gov*, n.d. www.youth.gov.

Alison Grace Johansen, "What Is a Router, and How Does It Work?," *Norton*, June 13, 2023. www.us.norton.com.

"What Is Telecommunication? (Definition and Types)," *Indeed*, June 24, 2022. www.indeed.com.

WEBSITES

ApprenticeshipUSA
www.apprenticeship.gov

ApprenticeshipUSA gives information about apprenticeship programs, including telecommunications technician programs.

CareerOneStop
www.careeronestop.org

CareerOneStop provides information and resources for people seeking jobs. It is sponsored by the US Department of Labor.

National Telecommunications and Information Administration
www.ntia.doc.gov

The National Telecommunications and Information Administration (NTIA) is an agency in the US government that handles matters of telecommunications and information.

INDEX

IMAGE CREDITS

Cover: © bunyiam/Shutterstock Images
5: © Henryk Sadura/Shutterstock Images
7: © Kosoff/Shutterstock Images
8: © Noiz Stocker/Shutterstock Images
10: © Vitpho/Shutterstock Images
13: © Anneka/Shutterstock Images
14: © Andrea Piacquadio/Pexels
17: © chaitawat/Pixabay
18: © Pavel Aloshin/Shutterstock Images
20: © Mircea Moira/Shutterstock Images
22 (caller, call recipient): © Olga Strel/Shutterstock Images
22 (cell towers): © Jemastock/Shutterstock Images
22 (MSC): © ADE2013/Shutterstock Images
24: © USA-Reiseblogger/Pixabay
25: © Andrey_Popov/Shutterstock Images
27: © Phovoir/Shutterstock Images
29: © Dragon Images/Shutterstock Images
30: © IndustryViews/Shutterstock Images
33: © Focal Point/Shutterstock Images
35: © eakkachai/Shutterstock Images
37: © Aspen/Nappy
39: © Rejdan/Shutterstock Images
40: © Stokkete/Shutterstock Images
42: © Valmedia/Shutterstock Images
44: © Martinelle/Pixabay
45: © Kappri/Shutterstock Images
47: © RobertLanger/Pixabay
48: © P. A./Shutterstock Images
51: © Gregory Charles/Shutterstock Images
52: © Jacob Lund/Shutterstock Images
55: © Panuwat Phimpha/Shutterstock Images
57: © Chalermphon_Tiam/Shutterstock Images

ABOUT THE AUTHOR

Christopher Forest is a middle school teacher in Massachusetts. He enjoys writing books for all ages. He has written nonfiction and fiction stories, articles, and novels for adults and children. In his spare time, he enjoys watching sports, playing guitar, reading, and spending time outdoors.